Y0-CFQ-112

E 640.42 DOU
Doudna, Kelly, 1963-
Let's subtract money
Carver County Library

SandCastle
Dollars & Cents

Let's Subtract
Money

Kelly Doudna

Consulting Editor Monica Marx, M.A./Reading Specialist

ABDO
Publishing Company

Published by SandCastle™, an imprint of ABDO Publishing Company, 4940 Viking Drive, Edina, Minnesota 55435.

Copyright © 2003 by Abdo Consulting Group, Inc. International copyrights reserved in all countries. No part of this book may be reproduced in any form without written permission from the publisher. SandCastle™ is a trademark and logo of ABDO Publishing Company. Printed in the United States.

Credits
Edited by: Pam Price
Curriculum Coordinator: Nancy Tuminelly
Cover and Interior Design and Production: Mighty Media
Photo Credits: Comstock, Hemera, PhotoDisc

Library of Congress Cataloging-in-Publication Data

Doudna, Kelly, 1963-
   Let's subtract money / Kelly Doudna.
       p. cm. -- (Dollars & cents)
   Includes index.
   Summary: Shows how to use subtraction to find out how much money one has left after paying for various items and looks at different denominations of money, from a penny to a twenty-dollar bill.
   ISBN 1-57765-901-5
   1. Money--Juvenile literature. 2. Subtraction--Juvenile literature. [1. Money. 2. Subtraction.] I. Title. II. Series.

HG221.5 .D657 2002
640'.42--dc21

2002071190

SandCastle™ books are created by a professional team of educators, reading specialists, and content developers around five essential components that include phonemic awareness, phonics, vocabulary, text comprehension, and fluency. All books are written, reviewed, and leveled for guided reading, early intervention reading, and Accelerated Reader® programs and designed for use in shared, guided, and independent reading and writing activities to support a balanced approach to literacy instruction.

Let Us Know

After reading the book, SandCastle would like you to tell us your stories about reading. What is your favorite page? Was there something hard that you needed help with? Share the ups and downs of learning to read. We want to hear from you! To get posted on the ABDO Publishing Company Web site, send us email at:

sandcastle@abdopub.com

SandCastle Level: Transitional

Coins and bills are money.

LIBERTY
IN GOD WE TRUST
1992
1972
FEDERAL RESERVE NOTE
THE UNITED STATES OF AMERICA
B 66067152 E
ONE DOLLAR
BC 92225640 A
FIVE DOLLARS
BF 61277637 A
TEN DOLLARS
BG 88724297 D
TWENTY DOLLARS
AG 89682037 B
FIFTY DOLLARS
AB 24634292 R
ONE HUNDRED DOLLARS

We use coins and bills
to pay for things.

Let's see what we can buy.

3¢

Max has 9 pennies.

The candy costs 3¢.
3¢ = 3 pennies

How many pennies will he have left?

Let's subtract.
9 - 3 = 6

45¢

The banana costs 45¢.
45¢ = 9 nickels

John has 3 nickels.

How many more nickels does he need?

Let's subtract.
9 - 3 = 6

$3.00

Lin has 8 one-dollar bills.

The kite costs $3.00.
$3.00 = 3 one-dollar bills

How many one-dollar bills will she have left?

Let's subtract.
8 - 3 = 5

80¢

The cookie costs 80¢.
80¢ = 8 dimes

Pat has 3 dimes.

How many more dimes does he need?

Let's subtract.
8 - 3 = 5

$15.00

Liz has 7 five-dollar bills.

The clock costs $15.00.
$15.00 = 3 five-dollar bills

How many five-dollar bills will she have left?

Let's subtract.
7 - 3 = 4

$1.75

The pinwheel costs $1.75.
$1.75 = 7 quarters

Brenda has 3 quarters.

How many more quarters does she need?

Let's subtract.
7 - 3 = 4

$60.00

Sue has 6 twenty-dollar bills.

The skates cost $60.00.
$60.00 = 3 twenty-dollar bills

How many twenty-dollar bills will she have left?

Let's subtract.
6 - 3 = 3

IN GOD WE TRUST
LIBERTY

IN GOD WE TRUST
LIBERTY

LIBERTY
IN GOD WE TRUST

LIBERTY
IN GOD WE TRUST
1992

FEDERAL RESERVE NOTE
THE UNITED STATES OF AMERICA
B 66067152 E
WASHINGTON, D.C.
ONE
B 66067152 E
ONE DOLLAR

FEDERAL RESERVE NOTE
BC 92252640 A
THE UNITED STATES OF AMERICA
FIVE
BC 92252640 A
FIVE DOLLARS

FEDERAL RESERVE NOTE
BG 88724297 D
THE UNITED STATES OF AMERICA
BG 88724297 D
TWENTY DOLLARS

What are these coins and bills called?

How much are they worth?

one penny = 1¢
one nickel = 5¢
one dime = 10¢
one quarter = 25¢
one dollar = $1.00
five dollars = $5.00
twenty dollars = $20.00

Index

banana, p. 9
bills, pp. 3, 5, 21
candy, p. 7
clock, p. 15
coins, pp. 3, 5, 21
cookies, p. 13
dime, p. 21
dimes, p. 13
dollar, p. 21
dollars, p. 21
five-dollar bills,
  p. 15
kite, p. 11

money, p. 3
nickel, p. 21
nickels, p. 9
one-dollar bills,
  p. 11
pennies, p. 7
penny, p. 21
pinwheel, p. 17
quarter, p. 21
quarters, p. 17
skates, p. 19
twenty-dollar bills,
  p. 19

Glossary

clock   a device used to measure time

kite   a lightweight frame covered with paper or another light material that is flown at the end of a long string

pinwheel   a toy with lightweight vanes that are pinned to a stick so they will spin in the wind

skates   boots with blades or wheels on the bottom

About SandCastle™

A professional team of educators, reading specialists, and content developers created the SandCastle™ series to support young readers as they develop reading skills and strategies and increase their general knowledge. The SandCastle™ series has four levels that correspond to early literacy development in young children. The levels are provided to help teachers and parents select the appropriate books for young readers.

Emerging Readers
(no flags)

Beginning Readers
(1 flag)

Transitional Readers
(2 flags)

Fluent Readers
(3 flags)

These levels are meant only as a guide. All levels are subject to change.

ABDO Publishing Company

To see a complete list of SandCastle™ books and other nonfiction titles from ABDO Publishing Company, visit www.abdopub.com or contact us at:
4940 Viking Drive, Edina, Minnesota 55435 • 1-800-800-1312 • fax: 1-952-831-1632